SUSTAINABLE LIVING UNPLUGGED

Embracing the Verdant Energy Revolution

TABLE OF CONTENTS

d. Acquiring Permits and Approvals

e. Securing Financing and Incentives

f. Solar Panel Installation

g. Electrifying and Grid Connection

h. Inspection and Testing

i. Monitoring and Maintenance

j. Revel in Solar Savings

k. Educate Yourself

CHAPTER 3: CAPTIVATING THE ZEPHYRS

- Unleashing Pristine Power: The Wind Energy Revolution
  - Prelude: Seizing the Murmurs of the Wind
- The Ballet of Wind and Turbines: How Wind Energy Operates
- The Potential of Wind Energy: From Zephyrs to Megawatts
- Communities in Motion: The Socioeconomic Impact of Wind Energy
- Evaluating Wind Energy Feasibility: Navigating Wind Patterns and Beyond
- Evolving with the Wind: Breakthroughs in Wind Energy Technology

CHAPTER 4: TAPPING INTO HYDROPOWER

- Unleashing the Torrent: The Enchantment of Flowing Water
  - Introduction: The Gracefulness of Streaming Water
- The Mechanics of Hydropower: Transforming Water's Energy into Power
- Water's Potential: The Environmental and Economic Gains of Hydropower
- Hydropower across Scales: From Small Streams to Grand Installations
- Navigating Hydropower Feasibility: Evaluating the Flow and Impact
- The Future Flow: Innovations and Transformations in Hydropower
- Streaming Towards a Sustainable Future

CHAPTER 5: GEOTHERMAL ENERGY: UNLEASHING EARTH'S HEAT

- Tapping Earth's Inner Fire: The Power of Geothermal Energy
  - Introduction: The Earth's Hidden Reservoirs of Energy

INTRODUCTION TO GREEN ENERGY

Green energy solutions refer to sources of energy that have a minimal impact on the environment and are considered renewable, meaning they can be naturally replenished over time. These solutions are a critical component of sustainable living, as they help reduce greenhouse gas emissions, dependence on fossil fuels, and other harmful environmental effects associated with traditional energy sources. Here are some common green energy solutions:

1. Solar Power: Solar panels convert sunlight into electricity. They can be installed on rooftops, in solar farms, and even integrated into clothing and accessories.

2. Wind Power: Wind turbines capture the kinetic energy of wind and convert it into electricity. Wind farms are often set up in areas with consistent wind patterns.

3. Hydropower: Hydropower generates electricity by harnessing the energy of flowing water, such as rivers

or waterfalls, to turn turbines. It's one of the oldest and most widely used renewable energy sources.

4. <u>Geothermal Energy</u>: Geothermal power taps into the Earth's internal heat to generate electricity or heat buildings directly. It's most viable in regions with high geothermal activity.

5. <u>Biomass Energy</u>: Biomass refers to organic materials like wood, agricultural residues, and even certain types of waste. Biomass can be burned to produce heat or converted into biofuels.

6. <u>Tidal and Wave Energy</u>: Tidal and wave energy harness the energy from ocean tides and waves to generate electricity. This technology is still developing and has the potential to be a reliable source of energy.

7. <u>Hybrid Systems</u>: Combining different renewable energy sources, such as solar and wind, can provide a more consistent and reliable energy supply.

8. <u>Microgrids and Smart Grids</u>: These technologies optimize the distribution and consumption of energy, enabling efficient integration of renewable sources and reducing wastage.

9. <u>Energy Storage</u>: Battery systems and other energy storage solutions help store excess energy generated by renewable sources for use during times of low energy production.

10. <u>Community Solar Programs</u>: These programs allow individuals who can't install solar panels on their own property to purchase a share of a larger solar installation and benefit from the generated energy.

11. <u>Green Building Design</u>: Incorporating energy-efficient design principles, passive solar techniques, and insulation can significantly reduce the energy consumption of buildings.

12. <u>Electric Vehicles (EVs) and charging Infrastructure</u>: Transitioning to electric vehicles and utilizing clean energy sources for charging further reduces carbon emissions from transportation.

13. <u>Off-Grid Solutions</u>: In remote areas or places with unreliable energy infrastructure, off-grid systems powered by renewable sources can provide consistent and sustainable energy.

14. <u>Decentralized Energy Generation</u>: Rather than relying solely on large power plants, distributed generation using renewable sources can reduce transmission losses and increase resilience.

Green energy solutions play a crucial role in reducing the carbon footprint and mitigating climate change. They contribute to a more sustainable and environmentally responsible way of meeting our energy needs. However, it's important to consider factors such as location, resource availability, and technological advancements when implementing these solutions.

CHAPTER 1

UNRAVELING THE DEMAND FOR GREEN ENERGY

In an era characterized by perplexing environmental conundrums and vexing climate concerns, the imperious mandate to metamorphose from archaic energy sources to sustainable alternatives is more imperative than ever. This chapter plunges into the pivotal role that verdant energy solutions perform in assuaging the repercussions of climate change and fostering a more serendipitous relationship between humanity and the planet.

The Ecological Toll of Conventional Energy

Conventional energy sources, notably fossil fuels like coal, oil, and natural gas, have historically propelled global development. However, their utilization takes a prodigious toll on the environment. The incineration of fossil fuels releases copious amounts of greenhouse gases, primarily carbon dioxide, into the atmosphere, propelling global warming, surging sea levels, and unleashing inclement weather phenomena. Moreover, the extraction and transportation of these fuels inflict irrevocable harm on ecosystems, befoul air and water, and jeopardize the delicate equilibrium of Earth's systems.

Green Energy's Promise of Emission Reduction

Green energy alternatives, anchored in renewable sources such as solar, wind, hydropower, and geothermal energy, present an efficacious antidote to this environmental quagmire. Unlike fossil fuels, these sources engender energy with minimal to zero direct emissions, ensuring the disruption of the pollution and climate cycle. By harnessing the might of natural elements, we can transmute away from fossil fuel reliance and significantly truncate our carbon footprint.

Renewable Energy's Symbiotic Relationship with Sustainability

The symbiosis between green energy and sustainability is profound. Renewable energy sources derive power from perpetual natural processes, ensuring an enduring bestowal of energy without depleting finite resources. This interconnectedness aligns flawlessly with the principles of sustainability, as it conserves the tenuous equilibrium of

Earth's ecosystems, fortifies energy security, and diminishes reliance on fossil fuels that jeopardize both the environment and future generations.

A Call to Collaborative Action

Understanding the imperativeness of green energy is an impassioned summons to collective action. As custodians of the Earth, we possess an ethical responsibility to metamorphose into sustainable energy sources. Whether as individuals, communities, or nations, our choices wield the power to transfigure our energy landscape and chart a trajectory towards a more sustainable future. By embracing green energy alternatives, we bridge the abyss between environmental consciousness and practical metamorphosis, fostering a cleaner, healthier, and more exuberant world for all.

CHAPTER 2

EMBRACING THE BRILLIANCE OF SOLAR ENERGY

Captivating the Sun's Power for an Ecologically Sound Future

Introduction: The Radiant Pledge of Solar Energy

This chapter unveils the metamorphic potential of solar energy, a guiding light in the realm of eco-friendly energy solutions. As we delve into the mechanics and advantages of solar power, you'll unearth how harnessing the limitless energy of the sun can revolutionize the way we energize our lives while mitigating the environmental repercussions of conventional energy sources.

Demystifying Solar Power: From Photons to Electricity

At the core of solar energy lies an extraordinary process: the metamorphosis of sunlight into electricity through photovoltaic (PV) panels. This section guides you through the intricate ballet of photons and electrons, elucidating how solar cells transmute sunlight into a pristine and renewable energy source. By unraveling the science behind solar panels, you'll cultivate a deeper reverence for their gracefulness and efficiency.

The Sun's Vow: Boons of Solar Energy

The advantages of solar energy extend far beyond its environmental merits. This section explores the economic and societal boons of embracing solar power. From diminished electricity bills and potential income generation through surplus energy sales to job creation and energy independence, the adoption of solar energy harbors a treasure trove of advantages for individuals, communities, and economies alike.

Illuminating Homes and Businesses: Solar Energy in Motion

Embarking on a solar odyssey encompasses more than just science—it's about tangible impact. Learn how solar energy systems can be seamlessly assimilated into residential and commercial settings, energizing homes, offices, factories, and even entire communities. Through case studies and triumph tales, you'll witness the palpable difference solar power can make in various applications.

Taking the Leap: Steps to Solar Installation

Transforming your solar aspirations into tangible reality demands a roadmap. This section provides a comprehensive guide to the steps required in installing solar panels on your property. From evaluating your energy needs and selecting the appropriate system size to navigating regulatory processes and choosing reputable installers, you'll acquire the insights necessary to embark confidently on your own to install odyssey.

There are various phases involved in **installing a solar energy system** to make sure you can use the sun's energy to create electricity for your house or place of business. The general stages for solar installation are as follows:

1. **Commence Evaluation and Strategizing:**
   - Ascertain your energy demands: Calculate your prevailing electricity consumption and discern your energy aspirations.

- Conduct a property inspection: Evaluate your premises' solar potential by contemplating factors like roof alignment, shading, and available area.

- Financial planning: Determine your budget for the solar installation, taking into consideration potential incentives and financing alternatives.

2. Select a Solar Engineer:

- Investigate and reach out to various solar installation firms.

- Obtain quotations and juxtapose them based on cost, warranties, reputation, and system architecture.

- Validate their credentials, licenses, and insurance coverage.

3. Conceptualizing the System:

- Collaborate with the chosen solar engineer to devise a system that fulfills your energy demands and takes into account your property's particulars.

- Decide on the type of solar panels, inverters, and other components to be employed.

4. Acquiring Permits and Approvals:

- Procure the necessary permits and endorsements from local authorities and utilities.
- Your solar engineer can often assist with this bureaucratic process.

5. Securing Financing and Incentives:

- Establish financial backing if required, such as loans or solar leases.
- Delve into and apply for any accessible solar incentives, tax credits, or rebates at the federal, state, or local level.

6. Solar Panel Installation:

- Your designated installer will arrange a date for the installation.
- Installers will typically affix solar panels on your roof or on the ground using racking systems.
- Connect panels to inverters, which transmute DC electricity to AC electricity.

7. Electrifying and Grid Connection:

- Unleash the might of your solar system as it intertwines with your electrical panel, empowering your home or business.
- If necessary, collaborate with your utility company to secure the enchantment of net metering or interconnection agreements.

8. Inspection and Testing:

- Prepare for the arrival of your local authorities or the utility company, who shall embark on a quest to ascertain that your system adheres to safety and regulatory standards.
- The installer, like a seasoned sorcerer, shall conjure tests to ensure the system operates flawlessly.

9. Monitoring and Maintenance:

- Weave a web of monitoring systems to trace the performance and energy production of your solar panels.
- Perform enchanting rituals of routine maintenance to ensure the system dances with optimal efficiency.

10. Revel in Solar Savings:

- Once your system receives the seal of approval and springs to life, embrace the bountiful rewards of solar energy, such as diminished electricity bills and environmental blessings.

11. Educate Yourself:

- Embark on a quest to decode the language of your solar system's monitoring data, as it unveils the secrets of its performance and savings.
- Delve into the depths of your system's warranties and maintenance requirements, and emerge enlightened.

12. Share the Experience:

- Consider bestowing the gift of your positive solar experience upon others, igniting a spark of renewable energy adoption within your community.

Always keep in mind that the specific steps and requirements for solar installation can vary depending on your location, local regulations, and the complexity of your project. It is therefore vital to forge a close alliance with a qualified solar installer and embark on a journey of thorough research, ensuring a triumphant and cost-effective solar installation.

Solar Energy's Evolution: Innovations and Future Horizons

The solar energy vista is perpetually evolving, with technological advancements and ingenious designs propelling its growth. Discover the latest breakthroughs in solar panel efficiency, energy storage solutions, and emerging trends such as building-integrated photovoltaics (BIPV) and solar-powered mobility. As the sun continues its ascent on the renewable energy frontier, you'll be equipped to embrace the future with enlightened optimism.

CHAPTER 3

CAPTIVATING THE ZEPHYRS

Unleashing Pristine Power: The Wind Energy Revolution

Prelude: Seizing the Murmurs of the Wind

In the dynamic realm of wind energy, the tender stroke of the wind metamorphoses into a formidable catalyst for transformation. From the mechanisms of wind turbines to the boundless potential of wind power, this chapter delves into how harnessing the kinetic energy of the wind can reshape our energy landscape while diminishing our carbon footprint.

The Ballet of Wind and Turbines: How Wind Energy Operates

At the core of wind energy lies the graceful interplay between wind and turbines. This segment delves into the mechanics of wind turbines, elucidating how their blades capture the kinetic energy of the wind and metamorphose it into electricity. As you plunge into the intricacies of wind energy generation, you'll develop an intense admiration for the engineering marvel that is the wind turbine.

The Potential of Wind Energy: From Zephyrs to Megawatts

Unlock the astonishing potential of wind power and its ability to reshape energy production. Explore how wind farms, frequently situated in locations with consistent wind patterns such as coastlines and vast plains, possess the capability to generate substantial amounts of clean electricity. Discover how harnessing the might of the wind can dramatically reduce greenhouse gas emissions and pave the way for a more sustainable energy future.

Communities in Motion: The Socioeconomic Impact of Wind Energy

Wind energy is not solely a technological phenomenon—it acts as a catalyst for positive transformation. This segment delves into the broader repercussions of wind power, including job creation, local economic growth, and community empowerment. Through real-world illustrations, you'll witness how wind energy projects have the potential to rejuvenate regions while promoting the adoption of clean energy.

Evaluating Wind Energy Feasibility: Navigating Wind Patterns and Beyond

Before the blades of wind turbines can whirl, a meticulous evaluation of wind patterns and environmental factors is imperative. This segment offers insights into assessing the feasibility of wind energy projects, encompassing wind resource assessment, site selection, and potential obstacles. Equipped with this knowledge, you'll be better prepared to assess the viability of wind energy in your vicinity.

Evolving with the Wind: Breakthroughs in Wind Energy Technology

The wind energy landscape perpetually evolves, propelled by technological advancements and innovation. Discover cutting-edge developments such as offshore wind farms, avant-garde turbine designs, and the integration of energy storage. By delving into these innovations, you'll grasp the transformative potential of wind energy in shaping our energy future.

From the scientific foundations to the socioeconomic impact and technological advancements, wind power emerges as a pivotal protagonist in our transition to cleaner energy sources.

Wind Energy

CHAPTER 4

TAPPING INTO HYDROPOWER

Unleashing the Torrent: The Enchantment of Flowing Water

Introduction: The Gracefulness of Streaming Water

Chapter 4 introduces you to plunge into the realm of hydropower, where the serene movement of water becomes a potent fountain of spotless and renewable energy. As we delve into the mechanics, advantages, and potential of hydropower, you'll unearth how this ancient practice has metamorphosed into a cornerstone of sustainable energy generation, fostering both environmental and societal concord.

The Mechanics of Hydropower: Transforming Water's Energy into Power

Unlock the mesmerizing process of how hydropower transmutes the kinetic energy of streaming water into electricity. This section delves into the technical intricacies of turbines, generators, and dams, unveiling the astonishing odyssey water undertakes from its natural flow to illuminating homes and industries. By comprehending the engineering marvel behind hydropower, you'll grasp its pivotal role in the renewable energy landscape.

Water's Potential: The Environmental and Economic Gains of Hydropower

Hydropower's potential unfurls far beyond its energy yield. This section examines the ecological benefits, such as curtailed greenhouse gas emissions and flood control, alongside the economic advantages that hydropower projects proffer to communities and nations. By embracing the concord between water resources and energy generation, you'll witness the

Hydropower across Scales: From Small Streams to Grand Installations

Discover the versatility of hydropower as it spans across scales, from micro-hydropower systems harnessing the energy of petite streams to colossal installations that energize entire regions. There is also showcasing real-world exemplars of hydropower's adaptability, providing insights into both the technical considerations and the societal benefits of diverse hydropower setups.

Navigating Hydropower Feasibility: Evaluating the Flow and Impact

Before embarking on a hydropower project, an exhaustive assessment of water resources and environmental impacts is indispensable. This section guides you through the intricacies of feasibility studies, environmental assessments, and regulatory considerations. Equipped with this knowledge, you'll be better prepared to evaluate the viability of hydropower initiatives in diverse settings.

The Future Flow: Innovations and Transformations in Hydropower

As technology advances, so does the potential of hydropower. Explore advancements such as run-of-river hydropower, pumped storage facilities, and breakthroughs in turbine design. By delving into these cutting-edge developments, you'll gain a glimpse into the future of hydropower, where renewable energy and environmental stewardship converge.

Conclusion: Streaming Towards a Sustainable Future

From the mechanical foundations to the multifaceted benefits and forward-looking innovations, hydropower emerges as a testament to humanity's ability to collaborate with nature for mutual progress. As we journey onward, guided by the rhythm of streaming water, we'll explore further green energy solutions that propel us towards a more harmonious and resilient world.

CHAPTER 5

GEOTHERMAL ENERGY: UNLEASHING EARTH'S HEAT

Tapping Earth's Inner Fire: The Power of Geothermal Energy

Introduction: The Earth's Hidden Reservoirs of Energy

Chapter 5 delves deep into the world of geothermal energy, an often-underestimated source of clean and reliable power derived from the Earth's internal heat. As we explore the science, benefits, and applications of geothermal energy, you'll uncover how tapping into the planet's innate warmth can revolutionize our approach to sustainable energy and environmental stewardship.

Geothermal Energy Unveiled: From Earth's Core to Practical Power

Unlock the remarkable journey of geothermal energy—from the Earth's molten core to the surface where it manifests as heat. This section delves into the mechanics of geothermal power generation, highlighting how heat from beneath the Earth's crust is harnessed to produce electricity. As you delve into the geological intricacies, you'll gain an appreciation for the intricacies of Earth's heat engine.

Harnessing Earth's Warmth: Benefits of Geothermal Energy

Beyond its scientific elegance, geothermal energy offers an array of benefits. Explore its potential to deliver clean, renewable, and consistent power. This section delves into the environmental advantages, from minimal greenhouse gas emissions to reduced reliance on fossil fuels. Moreover, discover how geothermal energy can provide reliable baseload power, contributing to energy security.

Geothermal in Action: Powering Homes, Industries, and More

Witness the practical application of geothermal energy across various domains. Explore how geothermal power plants, geothermal heating and cooling systems, and direct-use applications are transforming energy landscapes. Through real-world examples, you'll see how geothermal energy can heat homes, power industries, and even cultivate sustainable agriculture.

Navigating Geothermal Feasibility: Geological Considerations and Opportunities

Before tapping into Earth's heat, understanding geological conditions is paramount. This section guides you through assessing geothermal feasibility, encompassing geological surveys, resource potential, and environmental considerations. Armed with this knowledge, you'll be better equipped to evaluate the viability of geothermal projects in diverse settings.

The Geothermal Renaissance: Innovations and Pathways Forward

As technology evolves, geothermal energy's potential expands. Explore emerging innovations like enhanced geothermal systems (EGS), which tap into deeper heat reservoirs, and advancements in geothermal drilling techniques. This section provides a glimpse into the evolving landscape of geothermal energy, where innovation and sustainability converge.

CHAPTER 6

BIOMASS ENERGY AND BEYOND

Harvesting Nature's Bounty: Biomass Energy's Role in Sustainability

Embracing the Power of Organic Matter: Introduction

In Chapter 6, the topic of biomass energy is explored. This type of energy is produced from organic resources such as waste products, wood, and agricultural wastes. You'll learn how utilizing nature's abundance may help create a greener and more sustainable energy future as we examine the workings, advantages, and potential difficulties of biomass energy.

Creating Power from Organic Matter: The Alchemy of Biomass Energy

Learn how biomass energy is transformed into heat, electricity, and biofuels using the building components found in nature. This section examines several biomass conversion processes, including combustion, gasification, and anaerobic digestion, emphasizing the numerous paths by which whereby biological material may be used to generate energy.

From Waste to Wealth: Economic and Environmental Benefits

Biomass energy has a tremendous ability to handle waste management issues in addition to its energy output. Examine its ability to provide a sustainable energy alternative while keeping organic waste out of landfills. This section highlights the benefits for the environment, from lowering methane emissions to encouraging sustainable land use methods. Investigate the financial advantages that biomass initiatives bring to local economies and industry.

Applications and Real-world Effects of Biomass

Explore the range of biomass energy uses, from industrial facilities to rural families. Learn how biofuels, biogas, and heating systems based on biomass are changing how energy is produced and used. Witness how biomass energy helps with energy access and employment creation via case studies.

Navigating Biomass Feasibility: Balancing Resources and Impacts

A comprehensive approach is necessary to evaluate the viability of biomass projects, as it is with any energy source. The issues of feedstock availability, environmental effects, and legal compliance are walked through in this section. You will be better able to make sustainable judgments if you are aware of the variables that affect the feasibility of biomass projects.

Biomass Evolution: New Approaches and Prospects

The world of biomass energy is still changing as a result of technology. Discover innovations include new waste-to-energy uses, algae-based biofuels, and sustainable land management techniques. You may learn more about how biomass energy is developing to suit the needs of a shifting energy landscape by looking into these advances.

Creating a Sustainable Energy Landscape,

The transformational potential of biomass energy is revealed in Chapter 6. Biomass energy emerges as a testament to nature's ability to regenerate, from its biological roots to its promise in waste management and beyond. We'll keep looking at green energy options as we forge ahead, guided by the knowledge of harnessing organic matter and moving toward a more peaceful and resilient society.

CHAPTER 7

INTEGRATING GREEN ENERGY SOLUTIONS

Balancing Energies: Creating a Sustainable Power Mix

Introduction: Fusing Forces for a Brighter Future

As we explore the concept of hybrid systems, energy storage, and smart grids, you'll discover how harmonizing renewable sources and innovative technologies can reshape our energy landscape, fostering a resilient and interconnected future.

Hybrid Systems: Merging Forces for Consistency

Unlock the potential of hybrid energy systems, where different renewable sources are combined to ensure a consistent and reliable energy supply. This section delves into the intricacies of hybrid setups, such as solar wind systems, wind-hydro systems, and beyond. By embracing hybrid solutions, you'll witness how diverse energy sources collaborate to create a seamless energy ecosystem.

Energy Storage: Balancing Production and Consumption

Energy storage is the linchpin of a sustainable energy grid. Explore the transformative role of batteries, pumped storage, and other storage technologies in optimizing renewable energy use. This section showcases how energy storage enhances grid stability, facilitates energy access, and ensures uninterrupted power supply, even when the sun isn't shining or the wind isn't blowing.

Smart Grids and Microgrids: Empowering Decentralized Energy

Navigate the landscape of smart grids and microgrids, where intelligence meets energy distribution. Discover how digital technologies enable efficient energy management, real-time monitoring, and load balancing. This section explores how microgrids empower communities to generate, store, and distribute energy locally, fostering resilience and reducing reliance on centralized power sources.

Towards a Decentralized Future: Empowering Energy Communities

The decentralization of energy generation is reshaping how we power our lives. This section delves into the concept of energy communities, where individuals and local entities collectively generate, share, and manage energy resources. Witness how grassroots initiatives, combined with technological advancements, are democratizing energy and promoting sustainability.

Innovation Beyond Borders: International Grid Connectivity

Think beyond local solutions and explore the concept of international grid connectivity. This section delves into projects that link energy resources across borders, tapping into diverse renewable sources to ensure stability and energy security. By embracing cross-border collaborations, you'll gain insights into how interconnectedness enhances sustainability on a global scale.

Fusing Energies for a Resilient Tomorrow

Chapter 7 underscores the transformative power of integrating green energy solutions. From hybrid systems to smart grids, the synergy of renewables and innovative technologies holds the key to a resilient and interconnected energy future. As we progress, guided by the harmonization of energies, we'll continue to explore further green energy solutions that guide us toward a more sustainable and harmonious world.

CHAPTER 8

TRANSITIONING TO A GREENER LIFESTYLE

Shaping a Sustainable World: Empowering Individuals for Change

Introduction: Illuminating the Path to Sustainable Living

This chapter delves into the heart of sustainable living, exploring how individual choices can profoundly impact the environment and our collective future. As we journey through energy-efficient homes, eco-friendly habits, and conscious consumption, you'll discover how embracing a greener lifestyle can harmonize with green energy solutions, fostering a more sustainable and vibrant world.

Energy-Efficient Homes: Designing for Sustainability

Unveil the blueprint for energy-efficient homes, where design principles blend seamlessly with green technologies. This section explores passive solar design, energy-efficient appliances, and sustainable building materials. By integrating these elements, you'll uncover how homes can become beacons of eco-conscious living, reducing energy consumption and minimizing environmental impact.

Small Steps, Big Impact: Eco-Friendly Habits

Witness the transformative power of everyday choices. This section delves into eco-friendly habits that, when adopted collectively, can have a profound impact on reducing resource consumption and waste generation. Explore practices such as reducing single-use plastics, conserving water, and embracing plant-based diets, all of which contribute to a greener and more sustainable lifestyle.

Conscious Consumption: The Power of Mindful Choices

Shift the spotlight to consumption patterns and their far-reaching effects. This section encourages mindful decision-making, from purchasing products with sustainable certifications to supporting local and ethical businesses. By understanding the ripple effect of consumption, you'll see how individual choices can shape industries, influence supply chains, and drive positive change.

Transcending Transportation: Embracing Sustainable Mobility

The transportation sector is a crucial frontier for sustainability. This section explores the transition to electric vehicles (EVs), public transportation, biking, and walking. Discover how these choices reduce emissions, alleviate traffic congestion, and promote healthier lifestyles. By embracing sustainable mobility, you'll contribute to cleaner air and more livable cities.

Financial and Environmental Benefits: The Economics of Sustainable Living

Uncover the financial advantages of sustainable living. This section explores how energy efficiency, waste reduction, and eco-friendly choices can translate into long-term cost savings. By embracing solar panels, energy-efficient appliances, and greener transportation options, you'll witness how sustainability aligns with economic well-being.

The Power of Community: Collective Action for Change

Embrace the strength of community-driven initiatives. This section highlights the role of community gardens, local sustainability projects, and collaborative efforts in fostering sustainable living. By engaging with like-minded individuals, you'll experience the shared impact of collective action, demonstrating that small steps can lead to significant transformations.

Nurturing a Greener Tomorrow

We explore the essence of sustainable living as a harmonious dance between personal choices and global impact. From energy-efficient homes to conscious consumption, the greener lifestyle emerges as a catalyst for positive change. As we move forward, guided by mindful living, we'll continue to explore further green energy solutions that illuminate our path toward a more sustainable and regenerative world.

CHAPTER 9

GETTING AROUND OBSTACLES: EMPOWERING CHANGE IN THE FACE OF DIFFICULTIES

Introduction: Embracing Resilience Along the Way

Chapter 9 examines the obstacles that occur on the route to green energy adoption and sustainable living. This chapter provides you with insights and ways to overcome hurdles and sustain momentum, covering anything from misunderstandings to financial issues. You will be better equipped to contribute to a sustainable and resilient future if you recognize and solve these concerns.

Busting Myths: Dispelling Green Energy Myths

Discover the reality behind popular myths about green energy and sustainability. This section tackles concerns regarding intermittency, cost, and viability by giving evidence-based insights to dispel myths. By eliminating misconceptions, you'll obtain a better knowledge of green energy's advantages and possibilities..

Financial Considerations: Budgeting and Investing

Financial factors frequently impact judgments. This section investigates the early costs and long-term advantages of green energy adoption, from solar panel installation through appliance upgrades. Investigate financial incentives, tax credits, and financing solutions that might make environmentally friendly products more accessible and desirable.

Infrastructure and Regulatory Difficulties: Policy and Planning

Sustainable solutions frequently collide with policy and planning roadblocks. This section goes into the difficulties associated with grid integration, permits, and regulatory frameworks. Discover how communities and governments are collaborating to expedite procedures and foster an atmosphere conducive to green energy solutions.

Overcoming Change Resistance: Promoting Awareness and Action

Change resistance is a typical roadblock. This section looks at ways to break through inertia and inspire action. You'll discover how to accelerate collaborative efforts and urge people to adopt greener living, from community participation and education campaigns to establishing a culture of sustainability.

Making Holistic Decisions by Balancing Trade-Offs

Making sustainable decisions necessitates considering numerous elements. This section dives into the difficulties of decision-making at the intersection of environmental, social, and economic issues. Adopting a comprehensive approach can help you manage the trade-offs that come with seeking sustainable living and green energy solutions.

Advocacy for Empowerment: Creating Systemic Change

Individual actions can spark larger-scale change. This section looks at how advocacy and policy participation may help you have a bigger effect. Learn how people and organizations push for renewable energy objectives, sustainable transportation, and other structural reforms that help design a more sustainable future.

We empower ourselves to overcome hurdles and generate change by addressing myths, handling financial considerations, and inspiring collaborative action. We will investigate further green energy alternatives and contribute to a world that flourishes sustainably as we move forward, guided by resilience and resolve.

CONCLUSION

Your Contribution to a Greener Future

"Sustainable Living Unplugged: Switching to Green Energy Alternatives" has been a journey of knowledge, empowerment, and transformation for me. From recognizing the critical need for green energy solutions to adopting sustainable living habits, you've set out on a road that will not only benefit you but will help affect the future of our world.

You've learned about the physics and advantages of solar, wind, hydropower, geothermal, and biomass energy in these chapters. You've seen the power of combining various sources into coherent systems and seen how they can transform our energy landscape. You've explored a tapestry of possibilities that redefine our world, from harnessing the energy of the sun and wind to tapping into the Earth's heat and the abundance of biological matter.

You've also learned how these green energy solutions integrate with sustainable living habits. Energy-efficient houses, eco-friendly practices, and conscientious consumerism are no longer just buzz words; they are essential components of a lifestyle that is in sync with the requirements of our world. You've discovered that every decision, no matter how minor, contributes to a larger change toward a more sustainable and regenerative world.

"Sustainable Living Unplugged" is more than simply a collection of words; it's an invitation to change your life and make a difference in the world. With each conscientious decision, you advocate for a future in which clean energy, environmental stewardship, and vigorous life coexist. The road toward a greener future continues, led by your unshakable dedication to a world that flourishes in balance with nature.

Thank you for coming along with us on this adventure. Your decisions are important, and your capacity for good change is boundless.